Laboratory Notebook

Name _____

Email _____

Subject _____

Notebook # _____

Table of Contents

Page	Title	Date

Table of Contents

Page	Title	Date

Title		Date
Name	Partner	

Title		Date
Name	Partner	

2

Title		Date	
Name		Partner	

3

Title		Date
Name	Partner	

Title		Date
Name	Partner	

Title		Date
Name	Partner	

6

Title		Date
Name	Partner	

Title		Date
Name	Partner	

Title		Date	
Name		Partner	

9

Title		Date	
Name		Partner	

10

Title		Date	
Name		Partner	

Title		Date
Name	Partner	

Title		Date
Name	Partner	

Title		Date	
Name		Partner	

14

Title		Date
Name	Partner	

Title		Date
Name	Partner	

16

Title		Date
Name	Partner	

17

Title		Date	
Name		Partner	

Title		Date
Name	Partner	

19

Title		Date
Name	Partner	

Title		Date
Name	Partner	

Title		Date
Name	Partner	

22

Title		Date
Name	Partner	

Title		Date	
Name		Partner	

Title		Date
Name	Partner	

Title		Date
Name	Partner	

26

Title		Date	
Name		Partner	

27

Title		Date
Name	Partner	

Title		Date
Name	Partner	

29

Title		Date
Name	Partner	

30

Title		Date
Name	Partner	

Title		Date
Name	Partner	

Title		Date
Name	Partner	

33

Title		Date
Name	Partner	

Title		Date
Name	Partner	

Title		Date
Name	Partner	

Title		Date
Name	Partner	

Title		Date
Name	Partner	

Title		Date
Name	Partner	

Title		Date
Name	Partner	

Title		Date
Name	Partner	

41

Title		Date	
Name		Partner	

Title		Date
Name	Partner	

Title		Date
Name	Partner	

Title		Date
Name	Partner	

45

Title		Date
Name	Partner	

Title		Date
Name	Partner	

Title		Date
Name	Partner	

Title		Date	
Name		Partner	

Title		Date
Name	Partner	

Title		Date
Name	Partner	

Title		Date
Name	Partner	

53

Title		Date
Name	Partner	

Title		Date
Name	Partner	

Title		Date
Name	Partner	

56

Title		Date
Name	Partner	

57

Title		Date
Name	Partner	

Title		Date
Name	Partner	

Title		Date
Name	Partner	

Title		Date
Name	Partner	

62

Title		Date	
Name		Partner	

Title		Date
Name	Partner	

Title		Date
Name	Partner	

Title		Date
Name	Partner	

Title		Date
Name	Partner	

Title		Date
Name	Partner	

68

Title		Date	
Name		Partner	

Title		Date
Name	Partner	

Title		Date
Name	Partner	

Title		Date
Name	Partner	

Title		Date
Name	Partner	

73

Title		Date
Name	Partner	

Title		Date
Name	Partner	

Title		Date
Name	Partner	

Title		Date
Name	Partner	

Title		Date
Name	Partner	

78

Title		Date	
Name		Partner	

Title		Date
Name	Partner	

Title		Date
Name	Partner	

Title		Date
Name	Partner	

Title		Date
Name	Partner	

Title		Date	
Name		Partner	

Title		Date
Name	Partner	

Title		Date
Name	Partner	

Title		Date	
Name		Partner	

Title		Date
Name	Partner	

88

Title		Date
Name	Partner	

Title		Date
Name	Partner	

Title		Date	
Name		Partner	

Title		Date	
Name		Partner	

Title		Date
Name	Partner	

Title		Date
Name	Partner	

Title		Date
Name	Partner	

Title		Date
Name	Partner	

Title		Date	
Name		Partner	

Title		Date
Name	Partner	

Title		Date
Name	Partner	

Title		Date
Name	Partner	

Title		Date	
Name		Partner	

Title		Date
Name	Partner	

Title		Date
Name	Partner	

Title		Date
Name	Partner	

Title		Date
Name	Partner	

Title		Date	
Name		Partner	

106

Title		Date
Name	Partner	

Title		Date	
Name		Partner	

Title		Date
Name	Partner	

Title		Date
Name	Partner	

Title		Date
Name	Partner	

Title		Date
Name	Partner	

Title		Date
Name	Partner	

Title		Date	
Name		Partner	

Title		Date	
Name		Partner	

Title		Date
Name	Partner	

Title		Date	
Name		Partner	

117

Title		Date
Name	Partner	

119

Title		Date	
Name		Partner	

Title		Date
Name	Partner	

Title		Date
Name	Partner	

122

Title		Date
Name	Partner	

Title		Date
Name	Partner	

Title		Date	
Name		Partner	

125

Title		Date	
Name		Partner	

Title		Date
Name	Partner	

Title		Date
Name	Partner	

129

Title		Date
Name	Partner	

Title		Date
Name	Partner	

Title		Date
Name	Partner	

Title		Date
Name	Partner	

Title		Date
Name	Partner	

Title		Date
Name	Partner	

Title		Date	
Name		Partner	

Title		Date
Name	Partner	

Title		Date
Name	Partner	

Title		Date
Name	Partner	

144

Title		Date	
Name		Partner	

145

Title		Date
Name	Partner	

Title		Date
Name	Partner	

Made in the USA
Columbia, SC
22 August 2024

40992632R00085